Spend Some Time with 1 to 9

Mathematical Challenges for Increasing Number Sense and Fluency in Grades 6–12

by Dean Ballard

CORE, CORE Math, and Consortium on Reaching Excellence in Education are registered trademarks of Consortium on Reaching Excellence in Education, Inc.

Copyright © 2014 Consortium on Reaching Excellence in Education, Inc. All rights reserved. Printed in the United States of America. This publication is protected by copyright, and permission should be obtained from the publisher prior to any reproduction, storage in a retrieval system, or transmission in any form or by any means, electronic, mechanical, photocopying, recording, or otherwise.

ISBN: 978-0-9915355-1-4

For information about or to gain permission to use any content in this document, write to:
Permissions, Editorial Department
Consortium on Reaching Excellence in Education, Inc.
1300 Clay Street, Suite 600
Oakland, California 94612
Phone: (888) 249-6155
Fax: (510) 540-4242
Email: info@corelearn.com
www.corelearn.com

CORE Mission
CORE serves as a trusted advisor at all levels of preK–12 education, working collaboratively with educators to support literacy and math achievement growth for all students.

Our implementation support services and products help our customers build their own capacity for effective instruction by laying a foundation of research-based knowledge, supporting the use of proven tools, and developing leadership.

As an organization committed to integrity, excellence, and service, we believe that with informed school and district administrators, expert teaching, and well-implemented programs, all students can become proficient academically.

Table of Contents

Spend Some Time with 1 to 9, Grades 6–12

Challenge Number	Activity Title	Page Number
1	Make It 18 with 1 to 9	5
2	What Can You Do with 6 and 2?	7
3	What Can You Do with 5, 6, 8, and 2?	9
4	Create Variable Equations with Coefficients 1 to 9	11
5	Base, Height, and Area for Triangles with 1 to 9	13
6	Spend Unequal Fraction Time with 1 to 9	15
7	Multiply Your Fraction of Time with 1 to 9	17
8	Divide Your Fraction of Time with 1 to 9	19
9	All Around and Solid with 1 to 9	21
10	Create Equations with the Digits 1 to 9 and Their Opposites	23
11	Spend a Proportional Amount of Time with 1 to 9	25
12	Spend Proportional Time with Word Problems and 1 to 9	27
13	Spend Exponential Time with 1 to 9	29
14	Spend Some Radical Time with 1 to 9	31
15	Spend More Radical Time with 1 to 9	33
16	Make It Right with 1 to 9	35
17	Fill In a Linear Pattern with 1 to 9	37
18	Make a Point with Slope with 1 to 9	39
19	Make a Line with 1 to 9	41
20	Attack the Quadratic with 1 to 9	43
	Solutions to Challenges	45

© 2014 Consortium on Reaching Excellence in Education, Inc.

Introduction

Number sense, fluency, and problem solving are three important aspects of mathematics in all grades. Fluency with numbers, especially all operations with 1 through 9, is an essential foundation to build on year by year. However, it is not a foundation well built simply through flash cards. The foundation of fluency is built on the understanding of mathematical ideas and the connections between numbers.

In this booklet are sets of problems that build fluency and number sense through activities that require students to think about mathematical relationships. These problem-solving activities challenge students to reason about the mathematics at the same time as they are constantly putting numbers together and taking them apart.

Use these activities flexibly. Although a particular activity may revolve around a specific grade level, course, or standard, recognize its potential for use with other grade levels or courses. Use these activities as inspiration for other activities as you design your own versions. On the back side of each activity page are suggestions for prompts, questions, or extensions. Ask these and other questions of students to reveal their thinking and to engage in a process of making the mathematical connections explicit. Use extensions as opportunities for students to gain and apply deeper understanding of strategies they have just learned through the activity. The solutions for the activities are located in the back of the book. In many cases, a given solution is just one of many possible solutions. Recognize that students will often find other correct solutions.

Included with each activity are the corresponding Common Core State Standards for Mathematics (CCSSM). These are listed on the back side of each activity page. The activities also directly connect to the following Standards for Mathematical Practice in the CCSSM: 1) Make sense of problems and persevere in solving them; 2) Reason abstractly and quantitatively; 3) Construct viable arguments and critique the reasoning of others; 6) Attend to precision; and 7) Look for and make use of structure.

These activities are not a full curriculum or a lesson plan. They do not replace instruction on the important mathematical concepts embedded within the activities. Instead, these activities do provide additional tools to create a vigorous and engaging mathematics classroom and deepen students' understanding of essential math concepts and number relationships.

Suggestions for use:

- The activities are flexible enough to use with multiple grade levels and courses to help all students deepen their number sense. Expect the activities to take 15–45 minutes to complete.

- When using the extension questions and prompts provided at the back of each activity, you will need to add additional time.

- Have students first complete the online grades 6–12 Prep Time with 1 to 9 quiz related to the activity. (Turn to the next page for more information about the quizzes.)

- To facilitate a group or class of students completing an activity, you may reproduce the activity for handouts and/or place the activity page under a document camera to project it for others to see.

It is our hope that you will enjoy spending some time with 1 to 9!

Dean Ballard
Director of Mathematics
CORE, Inc.

Spend Some Time with 1 to 9, Grades 6–12

Prep Time with 1 to 9 Quizzes

The online Prep Time with 1 to 9 quizzes for grades 6–12 are designed to activate key prior knowledge and are provided to accompany each activity. The quizzes can be accessed through the CORE website using a computer, tablet, or smartphone. The Prep Time with 1 to 9 quizzes are a combination of information, questions, and hints that guide students through a step-by-step process to help prepare them to work on the related challenge activity in this book. Following is the CORE website link for access to the Prep Time with 1 to 9 quizzes:

http://www.corelearn.com/Products/Spend-Some-Time-with-1-to-9/

Challenge 1   **Spend Some Time with 1 to 9, Grades 6–12**

Make It 18 with 1 to 9

5

Use different digits from 1 to 9 to create a value of 18.

- For example, $4 + 6 + 8$, $9 \times (5 - 3)$, and $21 - (8 - 5)$ are correct because different digits were used within each expression, and each expression is equal to 18.

- However, $5 + 5 + 8$ is not correct because the digit 5 was used twice.

1. Create at least five different expressions with values of 18 that use three digits from 1 to 9.

2. Create at least five different expressions with values of 18 that use four and/or five digits from 1 to 9.

3. Create an expression with a value of 18 using as many of the digits from 1 to 9 as possible.

Spend Some Time with 1 to 9, Grades 6–12　　　　　　　　　　　　　　　　　　　**Challenge 1**

Make It 18 with 1 to 9

CCSSM: 1.OA.6, 2.OA.2
Prompts/Questions/Extensions

- Create an expression with a value of 18 using as many different operations as possible (e.g., use each of addition, subtraction, multiplication, and division at least once in the expression).

- Change the desired value to something like 15 or 20. This is especially useful after students share and discuss strategies with the original problem set and the mathematical connections are made explicit. Then students can try out newly learned strategies on a new set of numbers.

- Use negative and positive values (–9 to –1 and 1 to 9).

- Use radical values.

- Use values with decimals.

Challenge 2   **Spend Some Time with 1 to 9, Grades 6–12**

What Can You Do with 6 and 2?

1. Combine the digits 6 and 2 using any mathematical operation between or with them and/or just putting the numbers next to each other. Each digit, 6 and 2, must be used exactly once in each expression.

    a. What is the greatest possible value you can create?

    b. What is the least possible value you can create?

    c. How do you know you are correct for 1a and 1b above?

2. One way to "combine" two digits is to simply put them side by side with one digit in the ones place and the other digit in the tens place. For example, the digits 3 and 8 can be combined to create 38 (thirty-eight) or 83 (eighty-three). This is different than determining the product of the two digits, such as $(3)(8) = 24$.

    Is it always, sometimes, or never true that the product of any two positive digits, a and b, is less than either of the possible side-by-side arrangements of the two digits? That is,

    $(a)(b) < ab$ and $(a)(b) < ba$; for example, $(8)(3) < 83$, and $(8)(3) < 38$

    Justify or prove your answer.

© 2014 Consortium on Reaching Excellence in Education, Inc.

Spend Some Time with 1 to 9, Grades 6–12   **Challenge 2**

What Can You Do with 6 and 2?

CCSSM: 3.OA.4, 3.OA.5, 5.NBT.5-7, 7.NS.1

Prompts/Questions/Extensions

- Remind students, if appropriate, of the types of operations they can use, such as addition, subtraction, multiplication, division, exponents, square roots, fractions, just writing the digits next to each other, and so on. You could have the class brainstorm a list of possible operations.

- For a justification for question 2, look for discussion about the place value of the digits, and push for more than "It works with everything I have tried."

- Repeat questions 1a–1c with a different pair of numbers, such as 8 and 3, or add a third number to the mix, such as working with 6, 2, and 5.

8   © 2014 Consortium on Reaching Excellence in Education, Inc.

Challenge 3   **Spend Some Time with 1 to 9, Grades 6–12**

What Can You Do with 5, 6, 8, and 2?

Use the four digits 5, 6, 8, and 2 to make many other numbers. For example, the digits can be used to create 56 + 82, or 26 + 58, or 5 − 6 + 82, or 56 × 8 + 2. The rules are as follows:

- Use all four digits exactly once each time.

- Use any operation(s) and use at least one operation in each expression.

- Create only positive values as the overall value of the expression (values greater than zero).

1. What is the greatest positive integer value you can create?

2. What is the least positive integer value you can create?

3. What is the greatest possible positive rational number value you can create?

4. What is the least possible positive rational number value you can create?

5. For any of your answers for 1–4, explain or show how you know you have the greatest or least possible value that can be created.

Spend Some Time with 1 to 9, Grades 6–12 Challenge 3

What Can You Do with 5, 6, 8, and 2?

CCSSM: 1NBT.4, 2.OA.2, 2.NBT.5, 2.NBT.6, 5.NBT.5-7, 7.NS.1
Prompts/Questions/Extensions

- What strategies or reasoning did you use to get the greatest possible value or the least possible value?

- Limit the possible operations to addition, subtraction, multiplication, and division.

- Use a specific operation, such as exponents and/or radicals.

- Repeat questions 1–5 using a different set of numbers, such as 3, 4, 7, and 9. This is especially useful after students share and discuss strategies and the mathematical connections are made explicit. Then students can try out newly learned strategies on a new set of numbers.

Challenge 4   **Spend Some Time with 1 to 9, Grades 6–12**

Create Variable Equations with Coefficients 1 to 9

Create at least five variable equations with the following conditions:

- Use the digits 1–9 for coefficients.

- Use some or all of the digits in each equation.

- Do not use any digit more than once for the coefficients within any equation.

Examples:

$8x^2 \div 4x = 5x - 3x$ → uses the digits 3, 5, 4, and 8

$5x^2 - 4y + 6y + 3x^2 = 8x^2 + 2y$ → uses the digits 2, 3, 4, 5, 6, and 8

$(6x)(3x) = 18x^2$ → uses the digits 1, 3, 6, and 8

$$\frac{9y^5}{6y} = \frac{3}{2} y^4$$ → uses the digits 2, 3, 6, and 9

Nonexamples:

$(9x^2 \div 3x) + 6x = 9x$ → uses the digit 9 more than once

$10x = 6x + 4x$ → uses a digit that is not from 1 to 9, the digit 0

$8x^2 + 4x + 5 = 29 \times 1$ → is not a true equation

© 2014 Consortium on Reaching Excellence in Education, Inc.   **11**

Spend Some Time with 1 to 9, Grades 6–12   **Challenge 4**

Create Variable Equations with Coefficients 1 to 9

CCSSM: 6.EE.2-4, 7EE.1, A-APR.1, 7

Prompts/Questions/Extensions

- Create an equation that uses all nine digits as coefficients.

- Create at least one of each of the following types of equations: linear, quadratic, cubic, and exponential.

- Create a linear or quadratic equation such that the coefficients and the intercepts of the graph of the equation all use different digits from 1 to 9.

- What is the greatest number of terms you can have in one expression (on one side of the equal sign)?

- What is the greatest number of unlike variables you can have in one expression?

- Create an equation that uses as many different math operations as you can.

- Explain any strategies you used to create equations.

Base, Height, and Area for Triangles with 1 to 9

Triangles can be different sizes. Use the numbers 1 to 9 as the base and height of a triangle. The base and height must be different numbers.

1. Determine the areas of your triangles. Find at least seven triangles with areas such that the digits in the value for the area are only digits 1–9 (not zero) and are not the same as the digits used for the base and height of the triangle.

    You may choose to draw the triangles.

    For example:

    This triangle is correct because the base, height, and area all have different digits and include only the digits from 1 to 9.

    This triangle is not correct because the base and area both have the digit 4.

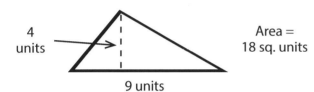

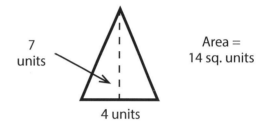

Spend Some Time with 1 to 9, Grades 6–12   **Challenge 5**

Base, Height and Area for Triangles with 1 to 9

CCSSM: 6.G.1

Prompts/Questions/Extensions

- Limit to a single-digit base and height, or if that was already the case, extend to using a multidigit base and/or height (e.g., $13 \times 4 =$ area of 26 sq. units).

- Explain why no combinations with 5 as the base or height will work for the area.

- Explain why no combinations with 2 as the base or height will work for the area.

- Prove that you have figured out all the possible triangles for area. (This may depend on the given constraints, such as whether or not a two-digit base or height is allowed.)

- Discuss strategies for determining triangles that work and for proving that all the possible triangles have been identified.

- Extension: Do this activity with trapezoids or parallelograms.

Challenge 6   Spend Some Time with 1 to 9, Grades 6–12

Spend Unequal Fraction Time with 1 to 9
Make the Inequality Statements True

1. Place **any** of the digits from the set above into the blank spaces in each set of inequality statements a–f shown to the right to make the statement true.

   For example, below we have used 1, 6, and 7 to make the inequality statement true:

   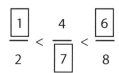

   - Do **not** use a digit more than once in the same statement.
   - Use only the digits in the set {1, 3, 5, 6, 7, 9}.
   - Create only **proper** fractions.

2. Show at least two possible solutions for any inequality statement that can have more than one solution.

3. Is there an inequality statement shown to the right that is not possible to solve? Justify or prove that it is not possible to solve.

4. What ideas or strategies did you use to help you solve some or all of these problems? Why do your ideas or strategies work?

a. $\dfrac{\square}{2} < \dfrac{4}{\square} < \dfrac{\square}{8}$

b. $\dfrac{\square}{2} < \dfrac{\square}{4} < \dfrac{8}{\square}$

c. $\dfrac{4}{\square} < \dfrac{\square}{2} < \dfrac{\square}{8}$

d. $\dfrac{\square}{4} < \dfrac{\square}{2} < \dfrac{8}{\square}$

e. $\dfrac{\square}{8} < \dfrac{\square}{2} < \dfrac{4}{\square}$

f. $\dfrac{8}{\square} < \dfrac{\square}{2} < \dfrac{\square}{4}$

© 2014 Consortium on Reaching Excellence in Education, Inc.

Spend Some Time with 1 to 9, Grades 6–12

Challenge 6

Spend Unequal Fraction Time with 1 to 9

CCSSM: 4.NF.2, 5.NF.2

Prompts/Questions/Extensions

- Discuss why 1/2 appears in every solution and how this affects students' thinking about the problem.

- What if the number set you had to work with was {1, 3, 5} for filling in the blanks? Show how you could complete some or all the inequality statements with just the numbers in this set. After students complete this task, discuss which ones can and cannot be completed and why.

- If you are allowed to create improper fractions, how would this change the answers for each set of inequality statements?

- If you are required to put the same number into each blank space for any of the inequality statements, in which statements would this work? In which statements would this be possible if you were also allowed to create improper fractions?

Challenge 7　　　　　　　　　　　　　　　　　　　　　　**Spend Some Time with 1 to 9, Grades 6–12**

Multiply Your Fraction Time with 1 to 9

Fill in each box with a digit from 1 to 9 so that the equation with fractions is true.

- No digit may be repeated within the entire equation.

For example:

This is correct because the filled-in digits make this equation true, and no digit is used more than once in the entire equation.

$$\frac{1}{2} \times \frac{3}{8} = \frac{\square - \square}{\square + \square + \square} \rightarrow \frac{1}{2} \times \frac{3}{8} = \frac{\boxed{9} - \boxed{6}}{\boxed{4} + \boxed{5} + \boxed{7}}$$

This is not correct because the digit 3 is used more than once in the equation.

$$\frac{1}{2} \times \frac{3}{8} = \frac{\square - \square}{\square + \square + \square} \rightarrow \frac{1}{2} \times \frac{3}{8} = \frac{\boxed{6} - \boxed{3}}{\boxed{4} + \boxed{5} + \boxed{7}}$$

Solve each equation below. If it is not possible to solve, explain why.

1. $$\frac{1}{2} \times \frac{3}{6} = \frac{\square - \square}{\square}$$

2. $$\frac{1}{2} \times \frac{6}{3} = \frac{\square + \square}{\square + \square}$$

3. $$\frac{\square}{\square} \times \frac{\square}{\square} = \frac{4 + 7 + 9}{5}$$

4. $$\frac{\square}{5} \times \frac{2}{\square} = \frac{1 + 7}{\square + \square}$$

5. $$\frac{\square}{\square} \times \frac{1}{4} = \frac{\square - \square}{8}$$

6. $$\frac{\square + \square}{8} \times \frac{4}{6} = \frac{7}{\square + \square}$$

7. $$\frac{\square + \square}{6} \times \frac{1}{3} = \frac{\square}{\square + \square}$$

© 2014 Consortium on Reaching Excellence in Education, Inc.　　　　　　　　　　　　　　**17**

Spend Some Time with 1 to 9, Grades 6–12

Challenge 7

Multiply Your Fraction Time with 1 to 9

CCSSM: 5.NF.1, 4

Prompts/Questions/Extensions

- What strategies did you use to solve the problems?

- Find other correct answers for questions 1 and 2.

- Create your own equivalent fraction multiplication problem.

- Create a fraction multiplication problem that uses as many digits as possible.

Challenge 8　　　　　　　　　　　　　　　　　　　　Spend Some Time with 1 to 9, Grades 6–12

Divide a Fraction of Time with 1 to 9

Fill in each box with a digit from 1 to 9 so that equivalent fractions are created.

- No digit may be repeated within the entire equation.

For example:

This is correct because no digit is used more than once in the entire equation, and the fractions are equal.

$$\frac{1}{2} \div \frac{7}{5} = \frac{\square - \square}{\square + \square} \rightarrow \frac{1}{2} \div \frac{7}{5} = \frac{9 - 4}{6 + 8}$$

This is not correct because the digits 1 and 6 are each used more than once in the equation.

$$\frac{1}{2} \div \frac{7}{5} = \frac{\square - \square}{\square + \square} \rightarrow \frac{1}{2} \div \frac{7}{5} = \frac{6 - 1}{6 + 8}$$

1. $\dfrac{2}{3} \div \dfrac{1}{4} = \dfrac{\square}{\square - \square}$

2. $\dfrac{2}{5} \div \dfrac{1}{3} = \dfrac{\square}{\square - \square}$

3. $\dfrac{2}{3} \div \dfrac{6}{7} = \dfrac{\square - \square}{\square + \square}$

4. $\dfrac{1}{5} \div \dfrac{3}{4} = \dfrac{\square - \square}{\square + \square}$

5. $\dfrac{\square - \square}{6} \div \dfrac{1}{4} = \dfrac{8}{3}$

6. $\dfrac{\square - \square}{4} \div \dfrac{3}{\square - \square} = \dfrac{1}{2}$

7. $1\dfrac{2}{3} \div \dfrac{5}{9} = \square - \square$

© 2014 Consortium on Reaching Excellence in Education, Inc.　　　　　　　　　　　　**19**

Spend Some Time with 1 to 9, Grades 6–12

Challenge 8

Divide a Fraction of Time with 1 to 9

CCSSM: 6.NS.1
Prompts/Questions/Extensions
• Create your own fraction division challenge problems.

All Around and Solid with 1 to 9
(Width, Height, Depth, and Surface Area)

Create rectangular solids by assigning the width, height, and depth of the solid using only the numbers 1 to 9, such that each dimension has a different length. Compute the surface area. Find all rectangular solids that have different digits in the width, height, depth, and surface area, and that only include the digits 1 to 9 (not zero).

You may choose to draw some or all of the rectangular solids.

For example:

The rectangular solid below is correct because the width, height, and depth, and surface area all have different digits.

The rectangular solid below is not correct because the surface area uses the digit 2 twice.

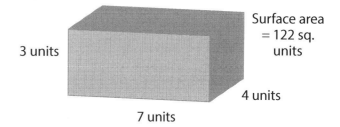

Spend Some Time with 1 to 9, Grades 6–12   **Challenge 9**

All Around and Solid with 1 to 9

CCSSM: 6.G.4, 7.G.6

Prompts/Questions/Extensions

Comparing volumes and surface areas of rectangular solids:

- Consider the rectangular solid with dimensions $1 \times 6 \times 8$ units that has a volume of 48 cubic units and a surface area of 124 sq. units. Compare this to the rectangular solid with dimensions $2 \times 4 \times 7$ units that has a volume of 56 cubic units and a surface area of 100 sq. units. The second solid has greater volume than the first but less surface area. Which box is smaller? Explain your answer.

- Identify at least three other pairs of rectangular solids such that as the volume increases, the surface area decreases. Explain why you think this occurs.

Challenge 10   **Spend Some Time with 1 to 9, Grades 6–12**

Create Equations with the Digits 1 to 9 and Their Opposites (–1 to –9)

Create as many equations as you can with the following conditions:

- Use the digits 1 to 9 and –1 to –9 to create many different equations.

- Use some or all of the digits in each equation.

- Do not use any digit more than once within any single equation.

- Do not use the digit 0.

- You may use any math operation, including exponents.

For example, $-8 + 1 = 5 - 3 + -9$ → uses the digits 1, 3, 5, –8, and –9

Extra Challenge: Create equations using the digits 1 to 9 and –1 to –9, but not 0, such that only the operations + and – are used on one side of the equation and only × and ÷ are used on the other side of the equation.

For example, $-8 + 1 + 5 = -6 \div 3$ → uses the digits 1, 3, 5, –8, and –6

© 2014 Consortium on Reaching Excellence in Education, Inc.   **23**

Spend Some Time with 1 to 9, Grades 6–12　　　　　　　　　　　　　　　　**Challenge 10**

Create Equations with the Digits 1 to 9 and Their Opposites

CCSSM: 6.NS.5

Prompts/Questions/Extensions

- Have students explain any of the strategies they used to create equations.

- Each side of an equation is called an *expression*. Can you change the value of any of your expressions by just adding or changing parentheses and still have a correct equation? Show this.

- What is the greatest value you can get on each side of an equation?

- What is the least value you can get on each side of an equation?

- Create an equation that uses all 18 digits.

- Create an equation that uses as many different math operations as you can.

Challenge 11   **Spend Some Time with 1 to 9, Grades 6–12**

Spend a Proportional Amount of Time with 1 to 9

$$y = \frac{a}{b}x$$

The equation above shows a proportional relationship between x and y.

(The constant of proportionality is $\frac{a}{b}$.)

Create a proportional relationship by using the digits 1 to 9 to fill in values for a, b, x, and y in the equation $y = \frac{a}{b}x$.

  - Create as many proportional equations that are true as you can.

  - Do not use a digit more than once in the same equation.

    Examples:

    - Correct answers:

      $2 = \frac{1}{4}(8)$ and $\frac{3}{2} = \frac{1}{4}(6)$ because in each equation no digit is used more than once and

      each equation is true.

    - Incorrect answers:

      $\frac{3}{2} = \frac{1}{4}(7)$ is not correct because the equation is not true.

      $3 = \frac{1}{3}(9)$ and $2\frac{2}{8} = \frac{1}{4}(9)$ are each not correct because the digit 3 is used twice in the first equation

      and the digit 2 is used twice in the second equation.

© 2014 Consortium on Reaching Excellence in Education, Inc.   **25**

Spend Some Time with 1 to 9, Grades 6–12 **Challenge 11**

Spend a Proportional Amount of Time with 1 to 9

CCSSM: 7.RP.2
Prompts/Questions/Extensions
• What are other ways to write the proportional relationship besides $y = \frac{a}{b}x$?
• How can other ways of writing the proportional relationship help you fill in correctly for a, b, x, and y?
• What if you were allowed to use zero as one of the digits? What other proportional relationships could you create?

Challenge 12

Spend Some Time with 1 to 9, Grades 6–12

Spend Proportional Time with Word Problems and 1 to 9

Change or create word problems such that the numbers in the problem and the solution all use different digits from 1 to 9.

For example:

Maggie runs 4 miles every 13 minutes. How far can she run in 26 minutes?

Answer: 8 miles (26 minutes is twice as much as 13 minutes, so she can run twice as far.)

The problem uses the digits 1, 2, 3, 4, and 6, and the answer uses the digit 8. Therefore, no digit is repeated in the problem and answer.

1. For each of the following word problems, change the numbers in the problems to create a new version for each problem such that the numbers in the new problem and in the solution use different digits from 1 to 9.

   a. Jack runs 4 miles every 13 minutes. How long will it take him to run 8 miles?

   b. Virginia spends $14 every 5 days on transportation to work. How much does she spend in 3 days?

2. Create at least one more version for each of the word problems above such that the numbers in the new problems and in the solutions use different digits from 1 to 9.

3. Create your own word proportional-problem such that the numbers in the problem and in the solution all use different digits from 1 to 9.

© 2014 Consortium on Reaching Excellence in Education, Inc.

27

Spend Some Time with 1 to 9, Grades 6–12　　　　　　　　　　　　　　　　　**Challenge 12**

Spend Proportional Time with Word Problems and 1 to 9

CCSSM: 7.RP.2
Prompts/Questions/Extensions
• What is the unit rate in each of the word problems on the previous page, including the ones you created yourself?
• Show two different ways to determine the solution for at least two of the problems shown on the previous page. (In this context, *solution* means the solution to the word problem, such as figuring out the solution of 8 miles in the example.)

28　　　　　　　　　　　　　　　　© 2014 Consortium on Reaching Excellence in Education, Inc.

Challenge 13   Spend Some Time with 1 to 9, Grades 6–12

Spend Exponential Time with 1 to 9
Fill In Each Inequality to Make It True

1. Place **any** of the five numbers from the set above into the blank spaces in each inequality shown to the right to make the statement true.

   Do not use a number more than once in any inequality statement.

   For example:

   $4^{\boxed{3}} < \boxed{9}^2 < 8^{\boxed{5}}$

   a. $4^{\square} < \square^2 < 8^{\square}$

   b. $\square^2 < 4^{\square} < 8^{\square}$

   c. $\square^2 < 8^{\square} < 4^{\square}$

2. Are there any statements that are impossible to make true? Why?

   d. $4^{\square} < 8^{\square} < \square^2$

   e. $8^{\square} < 4^{\square} < \square^2$

3. Show at least two possible solutions for any statement that can have more than one solution.

   f. $8^{\square} < \square^2 < 4^{\square}$

4. What ideas or strategies did you use to help you solve some or all of these problems? Why do your ideas or strategies work?

Spend Some Time with 1 to 9, Grades 6–12　　　　　　　　　　　　　　　**Challenge 13**

Spend Exponential Time with 1 to 9

CCSSM: 6.EE.1

Prompts/Questions/Extensions

- What if the number set you had to work with was {1, 3, 5}, and each number was to be used once in each inequality? Show how you could complete some or all the inequalities with just these three numbers.

- Change the given numbers in the inequalities from 2, 4, and 8 to 3, 6, and 9, respectively, and complete the inequalities choosing from the set {2, 4, 6, 8}. This activity is especially beneficial after students share and discuss strategies with the original problem set and the mathematical connections are made explicit. Then students can try out newly learned strategies on a new set of numbers.

Challenge 14 — Spend Some Time with 1 to 9, Grades 6–12

Spend Some Radical Time with 1 to 9
Make the Inequality Statements True

1. Place **any** of the digits from the set above into the blank spaces in each inequality shown to the right to make the statement true.

   For example, below we have used 3, 5, and 7 to make a true statement:

   $$\boxed{3}\sqrt{2} < \sqrt{4\boxed{5}} < \sqrt{\boxed{7}8}$$

   - Do not use a digit more than once in the same statement.
   - Do not use a calculator.

2. Show at least two possible solutions for any problem that can have more than one solution.

3. If you were required to place the same number in each blank, is there any statement that is impossible to solve with this condition? If so, explain or prove why there is no possible solution in these cases.

4. What ideas or strategies did you use to help you solve some or all of these problems? Why do your ideas or strategies work?

a. $\square\sqrt{2} < \sqrt{4\square} < \sqrt{\square 8}$

b. $\sqrt{4\square} < \square\sqrt{2} < \sqrt{\square 8}$

c. $\sqrt{\square 8} < \sqrt{4\square} < \square\sqrt{2}$

d. $\sqrt{\square 8} < \square\sqrt{2} < \sqrt{4\square}$

e. $\sqrt{4\square} < \sqrt{\square 8} < \square\sqrt{2}$

f. $\square\sqrt{2} < \sqrt{\square 8} < \sqrt{4\square}$

© 2014 Consortium on Reaching Excellence in Education, Inc.

Spend Some Time with 1 to 9, Grades 6–12   **Challenge 14**

Spend Some Radical Time with 1 to 9

CCSSM: 8.NS.1, 8.NS.2

Prompts/Questions/Extensions

- What if the number set you had to work with was {1, 3, 5} for filling in the boxes? Show how you could complete some or all the inequalities with just the numbers in this set. Discuss which ones can and cannot be completed and why.

- Change the given numbers in the inequalities from 2, 4, and 8 to 3, 6, and 9, respectively, and complete the inequalities choosing from the set {2, 4, 6, 8}. This activity is especially beneficial after students share and discuss strategies with the original problem set and the mathematical connections are made explicit. Then students can try out newly learned strategies on a new set of numbers.

Challenge 15   Spend Some Time with 1 to 9, Grades 6–12

Spend More Radical Time with 1 to 9
Make the Inequality Statements True

$$\boxed{1}\ \boxed{3}\ \boxed{5}\ \boxed{7}\ \boxed{9}$$

1. Place **any** of the digits from the set above into the blank spaces in each inequality shown to the right to make the statement true.

   For example, below we have used 3, 5, and 9 to make a true statement:

   $$\sqrt{\boxed{3}\,2} < 4\sqrt{\boxed{5}} < \sqrt{8\boxed{9}}$$

   • Do not use a digit more than once in the same statement.

   • Do not use a calculator.

2. Show at least two possible solutions for any problem that can have more than one solution.

3. Is there any statement that is impossible to solve? Prove there is no possible solution.

4. What ideas or strategies did you use to help you solve some or all of these problems? Why do your ideas or strategies work?

a. $\sqrt{\boxed{}\,2} < 4\sqrt{\boxed{}} < \sqrt{8\boxed{}}$

b. $4\sqrt{\boxed{}} < \sqrt{\boxed{}\,2} < \sqrt{8\boxed{}}$

c. $\sqrt{8\boxed{}} < 4\sqrt{\boxed{}} < \sqrt{\boxed{}\,2}$

d. $\sqrt{8\boxed{}} < \sqrt{\boxed{}\,2} < 4\sqrt{\boxed{}}$

e. $4\sqrt{\boxed{}} < \sqrt{8\boxed{}} < \sqrt{\boxed{}\,2}$

f. $\sqrt{2\boxed{}} < \sqrt{8\boxed{}} < 4\sqrt{\boxed{}}$

© 2014 Consortium on Reaching Excellence in Education, Inc.   **33**

Spend Some Time with 1 to 9, Grades 6–12   **Challenge 15**

Spend More Radical Time with 1 to 9

CCSSM: 8.NS.1, 8.NS.2

Prompts/Questions/Extensions

- What if the number set you had to work with was {1, 3, 5} for filling in the boxes? Show how you could complete some or all the inequalities with just the numbers in this set. Discuss which ones can and cannot be completed and why.

- Change the given numbers in the inequalities from 2, 4, and 8 to 3, 6, and 9, respectively, and complete the inequalities choosing from the set {2, 4, 6, 8}. This activity is especially beneficial after students share and discuss strategies with the original problem set and the mathematical connections are made explicit. Then students can try out newly learned strategies on a new set of numbers.

- If you were required to place the same number in each blank, is there any statement that is impossible to solve with this condition? If so, explain or prove why there is no possible solution in such a case.

Make It Right with 1 to 9

Create as many right triangles as you can by assigning the base and altitude using the numbers 1 to 9. Compute the value for the hypotenuse. Create triangles such that the values for the three sides all use different digits from 1 to 9 (zero is not allowed).

You may choose to draw the triangles.

For example:

The right triangle below is correct because no digit is used more than once in the values for the three sides.

The right triangle below is not correct because the base and the hypotenuse both have the digit 4.

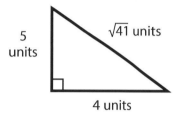

Spend Some Time with 1 to 9, Grades 6–12   **Challenge 16**

Make It Right with 1 to 9

CCSSM: 8.G.7

Prompts/Questions/Extensions

- Require radicals be in their simplest form.

- There are many correct answers for this problem. It would make a good partner or group assignment so that students could assign different sets of numbers to try. For example, if a group of three students—student A, student B, and student C—worked together on this, student A might try all right triangles with bases of 1, 2, or 3. Student B might try all right triangles with bases of 4, 5, or 6. Student C might try all right triangles with bases of 7, 8, or 9.

- Guide students to recognize that if they try a triangle with base = 3 units and altitude = 5 units, they do not need to also try the triangle with base = 5 units and altitude = 3 units, because these are congruent triangles.

- Have students prove they have found all possible triangles that work (with a single-digit integer base and altitude).

- Have students first assign the base and hypotenuse, and then compute the altitude.

- Repeat this activity with other shapes, such as parallelograms (assigning the base and altitude or the slant height), trapezoids (assigning the two bases and the altitude, or the two bases and the two slant heights), and circles (assigning the radius and using exact values for circumference rather than an approximation for pi).

Challenge 17   Spend Some Time with 1 to 9, Grades 6–12

Fill In a Linear Pattern with 1 to 9

Fill in the IN/OUT tables below with pairs of numbers that match the rules for the tables.

- There are three pairs of numbers for each table.

- No digit is used more than once in each table.

- Create two different tables for each rule.

For example:

The tables below are correct because there are three pairs of numbers that match the rule in each table, no digit is used more than once in each table, and the two tables are not exactly the same.

The tables below are not correct because the digit 2 is used twice in the second table.

IN	OUT
1	3
2	5
4	9

Rule:
OUT = (IN × 2) + 1

IN	OUT
2	5
3	7
4	9

Rule:
OUT = (IN × 2) + 1

IN	OUT
2	5
3	8
6	17

Rule:
OUT = (IN × 3) – 1

IN	OUT
1	2
2	5
3	8

Rule:
OUT = (IN × 3) – 1

IN	OUT

Rule:
OUT = (IN × 2) – 1

IN	OUT

Rule:
OUT = (IN × 2) – 1

IN	OUT

Rule:
OUT = (IN × 2) – 3

IN	OUT

Rule:
OUT = (IN × 2) – 3

IN	OUT

Rule:
OUT = (IN × 3) – 3

IN	OUT

Rule:
OUT = (IN × 3) – 3

IN	OUT

Rule:
OUT = (IN × 2) + 3

IN	OUT

Rule:
OUT = (IN × 2) + 3

© 2014 Consortium on Reaching Excellence in Education, Inc.

Spend Some Time with 1 to 9, Grades 6–12 **Challenge 17**

Fill In a Linear Pattern with 1 to 9

CCSSM: 4.OA.5, 5.OA.3

Prompts/Questions/Extensions

- What strategies did you use to figure out the numbers for the tables?

- Create two more IN/OUT tables for the rule (IN × 3) – 3 such that neither table is exactly like any other table you have created for this rule.

- Create two more IN/OUT tables for the rule (IN × 2) + 3 such that neither table is exactly like any other table you have created for this rule.

Challenge 18   **Spend Some Time with 1 to 9, Grades 6–12**

Make a Point of Slope with 1 to 9

Identify two points in quadrant 1 of the coordinate plane that a line passes through such that the slope of the line does not use any of the digits in the two points.

- For example, a line through the points (8, 7) and (5, 6) has slope of 1/3. The digits in the slope and the digits in the two points are all different. [Slope = (7 – 6)/(8 – 5) = 1/3.] Therefore, this is a correct solution for this challenge.

- However, a line through the points (7, 5) and (2, 1) has slope of 4/5. The digit or number 5 is used twice, once in the point (7, 5) and again in the slope. [Slope = (5 – 1)/(7 – 2) = 4/5.] Therefore, this is not a correct solution for this challenge.

1. Create at least seven correct solutions for this challenge.

2. Create at least two solutions that use at least one double-digit number.

   For example, (14, 5) and (8, 3) ➜ (5 – 3)/(14 – 8) = 2/6

3. Is it possible to have solutions that use only even or only odd numbers? Prove your answer.

© 2014 Consortium on Reaching Excellence in Education, Inc.   **39**

Spend Some Time with 1 to 9, Grades 6–12

Challenge 18

Make a Point of Slope with 1 to 9

CCSSM: 8.F.3, 8.F.4, HS.F-IF.4, HS.F-IF.9

Prompts/Questions/Extensions

- Is it possible to have two points such that both coordinates in both points have two-digit numbers? Why or why not?

- Create a solution that uses all nine digits.

- For any given solution, how many other solutions can be created using exactly the same digits?

Challenge 19   **Spend Some Time with 1 to 9, Grades 6–12**

Make a Line with 1 to 9

$$y = mx + b$$

The equation above is the slope-intercept form of a linear equation.

1. Create a pair of linear equations that represent parallel lines by filling in values for m and b in the equation $y = mx + b$.

   - No digit can be used twice in the pair of equations.

   - Use only the digits 1 to 9 (zero cannot be used in the equations).

     Examples:

     - The two equations $y = 2x + 5$ and $y = \dfrac{6}{3}x + 7$ are a correct solution because they represent parallel lines and no digit is used more than once in the set.

     - The two equations $y = 2x + 5$ and $y = 3x + 8$ are not correct because the lines represented by the equations are not parallel.

     - The two equations $y = \dfrac{2}{4}x + 5$ and $y = \dfrac{1}{2}x + 9$ are not correct because the digit 2 is used twice.

2. Create a pair of linear equations that represent perpendicular lines by filling in values for m and b in the equation $y = mx + b$.

   - No digit can be used twice in the pair of equations.

   - Use only the digits 1 to 9 (zero cannot be used in the equations).

© 2014 Consortium on Reaching Excellence in Education, Inc.   **41**

Spend Some Time with 1 to 9, Grades 6–12 **Challenge 19**

Make a Line with 1 to 9

CCSSM: 8.F.3, 8.F.4, HS.F-IF.4, HS.F-IF.9

Prompts/Questions/Extensions

- Why is $y = mx + b$ called the slope-intercept form of a linear equation?

- What does it mean for two lines to be parallel?

- What does it mean for two lines to be perpendicular?

- How can you tell that $y = 2x + 5$ and $y = \frac{6}{3}x + 7$ are parallel, and that $y = 2x + 5$ and $y = 3x + 8$ are not parallel?

- How can you tell from the equations that two lines are perpendicular?

- What strategies did you use to create pairs of parallel lines?

- What strategies did you use to create pairs of perpendicular lines?

Challenge 20   **Spend Some Time with 1 to 9, Grades 6–12**

Attack the Quadratic with 1 to 9

$$y = (x + m)(x + n) \quad \rightarrow \quad y = ax^2 + bx + c$$

On the left above is a quadratic equation written in factored form. The same equation can also be written in standard form, as shown on the right above, with $a = 1$. Two such equations create a pair of equivalent quadratic equations.

Create equivalent quadratic equations in factored and standard form by filling in numbers from 1 to 9 for m and n and then rewriting your equations in standard form. (Use only single-digit numbers for m and n.)

- Note: The standard form **is** permitted to have the digit 0 in it.

- **In this activity**, equivalent pairs of quadratic equations (factored form and standard form) cannot use any digit from 1 to 9 more than once in the pair of equations (the coefficient of 1 in front of x^2 does not count and the exponent 2 in x^2 does not count as part of the digits from 1 to 9).

For example:

The pair of equivalent quadratic equations below is correct because no digit is used more than once in the **pair** of equations.

$$y = (x + 4)(x + 3) \quad \rightarrow \quad y = x^2 + 7x + 12$$

The pair of equivalent quadratic equations below is not correct because the digit 2 is used in each equation.

$$y = (x + 1)(x + \mathbf{2}) \quad \rightarrow \quad y = x^2 + 3x + \mathbf{2}$$

Create as many pairs of equivalent quadratic equations as you can.

_____

_____

_____

_____

_____

_____

_____

_____

_____

_____

_____

_____

_____

_____

© 2014 Consortium on Reaching Excellence in Education, Inc.

Spend Some Time with 1 to 9, Grades 6–12 **Challenge 20**

Attack the Quadratic with 1 to 9

CCSSM: HS.A.REI.4b, HS.A-SSE.2, HS.A-SSE.3

Prompts/Questions/Extensions

- Create all the sets of equivalent quadratic equations that are possible. Prove that you have created all that are possible.

- Create two **sets** of equivalent quadratic equations such that no digit is used more than once within and between the two sets.

- Can all quadratic equations written in factored form also be written in standard form? Explain why or why not.

- Can all quadratic equations written in standard form also be written in factored form? Explain why or why not.

- Find a set of equivalent quadratic equations such that there is no 1 digit in the values for m, n, b, or c.

Solutions to
Spend Some Time with 1 to 9, Grades 6–12
Challenges

Spend Some Time with 1 to 9, Grades 6–12 **Challenge 1 Solution**

Make It 18 with 1 to 9

1. Sample solutions:

   $5 + 6 + 7$, $4 + 6 + 8$, $3 + 6 + 9$, $2 + 7 + 9$, $1 + 8 + 9$, $3 + 7 + 8$, $4 + 5 + 9$

   $16 + 2$, $15 + 3$, $25 - 7$, $36/2$, $1 \times 3 \times 6$, $3^2 + 9$, etc.

2. Sample solutions:

   $1 + 2 + 6 + 9$, $1 + 2 + 7 + 8$, $1 + 3 + 5 + 9$, $1 + 3 + 6 + 8$, $1 + 4 + 5 + 8$, $1 + 4 + 6 + 7$, $3^2 + 4 + 5$,

   $2 \times (1 + 3 + 5)$, $36 - 18$, $(54/3) \times (2 - 1)$, $1 + 2 + 3 + 4 + 8$, $1 + 2 + 3 + 5 + 7$, $1 + 2 + 4 + 5 + 6$,

   $3^2 + 7 + 6 - 4$, etc.

3. Sample solutions:

   $9 \times (8 - 7) \times (6 - 5) \times (4 - 3) \times (2 \times 1)$

   $(98/14) + 2 + 3 + 5 + 7 - 6$, etc.

Challenge 2 Solution   **Spend Some Time with 1 to 9, Grades 6–12**

What Can You Do with 6 and 2?

1a. The answer depends on the math level of the students. Either 64 or 62 may be acceptable. If factorials are included, then the answers become very large (however, few students are expected to understand how to use factorials).

$$62; \qquad 2^6 = 64; \qquad 62! = 3.147 \times 10^{85}; \qquad (2^6)! = 1.269 \times 10^{89}$$

1b. The answer depends on the math level of the students.

$$2 - 6 = -4; \qquad 2/6 = 1/3; \qquad 6 - 2 = 4; \qquad 6 \div 2 = 3; \qquad -(2^6)! = -1.269 \times 10^{89}$$

1c. Creating a number by putting the two digits next to each other will make the greatest value if neither exponents nor factorials are used. Trying several operations, including multiplication and addition, will lead to 62 (or 64 if exponents are used) as the greater value. Students should understand that multiplication usually creates a greater value than addition with whole numbers, but they should see that neither creates a greater value than just putting the digits next to each other ($62 > 2 \times 6$).

For the least value, division and subtraction create lesser values. If students are not yet working with negative values, then division generally results in a lesser value than subtraction with whole numbers.

2. It is always true.

- Students may only come to this conclusion based on observations from trying different numbers.

- However, to prove it, we need to use place value understanding. When we set the digits side by side—for example, ab—one of them will be in the tens place, making that digit's actual value $a \times 10$ in our example. So the value of ab is $(a \times 10) + b$. This is the expanded form of the place value number. However, if we multiply $a \times b$, this must be less than $(a \times 10) + b$ because $b < 10$: $a \times b < a \times 10 < (a \times 10) + b$.

© 2014 Consortium on Reaching Excellence in Education, Inc.

Spend Some Time with 1 to 9, Grades 6–12

Challenge 3 Solution

What Can You Do with 5, 6, 8, and 2?

1. Without exponents or factorials: $82 \times 65 = 5330$

    With exponents: 6^{852}, $6^{(8^{(5^2)})}$, or exponents with factorials, $6^{(852!)}$

2. $(6 + 5) - (8 + 2) = 1$

3. Same answers as 1.

4. Without exponents or factorials: 2/865

    With exponents and/or factorials: $2/(8^{65})$, $6^{(-852)}$, $6^{(-852!)}$

5. Answers will vary. For example, for the greatest and least integers:

    *   If only addition, subtraction, multiplication, and division are used, then all possible combinations with multiplication can be tried to prove that 5330 is the greatest possible value (this also requires explaining why multiplication is the only operation of the four that needs to be tested).

    *   If exponents are used, then a comparison of different bases and exponents will show that 6^{852} is the largest possible value, unless embedded exponents or exponents with factorials are used. Reasoning about exponents should lead to recognizing that the larger the exponent with relatively similar bases, the larger the overall value.

    *   The least integer: Since the lowest positive integer is 1, then creating an expression equivalent to 1 means we have created the lowest possible integer.

    *   The least rational number: Considering only positive rational numbers, then a unit fraction with the greatest possible rational number as the denominator will have the least value. If we include negative values, then the opposite of the greatest possible rational number will be the least possible rational number.

Challenge 4 Solution

Spend Some Time with 1 to 9, Grades 6–12

Create Variable Equations with Coefficients 1 to 9

There are many different possible equations.

- Here is an example of an equation with seven terms:

$$9x - 8x + 7x - 6x + 4x - 3x + 2x = 5x$$

- Here is an example of an equation that uses all nine digits:

$$9a - (6a + 1a) + 3a = \frac{5a^2}{(7a + 4a) - (8a + 2a)}$$

Spend Some Time with 1 to 9, Grades 6–12

Challenge 5 Solution

Base, Height, and Area for Triangles with 1 to 9

Shaded areas identify correct answers (using only one-digit base and height).

Base	Height	Area
1	2	1
1	3	1.5
1	4	2
1	5	2.5
1	6	3
1	7	3.5
1	8	4
1	9	4.5
2	3	3
2	4	4
2	5	5
2	6	6
2	7	7
2	8	8
2	9	9

Base	Height	Area
3	4	6
3	5	7.5
3	6	9
3	7	10.5
3	8	12
3	9	13.5
4	5	10
4	6	12
4	7	14
4	8	16
4	9	18

Base	Height	Area
5	6	15
5	7	17.5
5	8	20
5	9	22.5
6	7	21
6	8	24
6	9	27
7	8	28
7	9	31.5
8	9	36

50

© 2014 Consortium on Reaching Excellence in Education, Inc.

Challenge 6 Solution

Spend Some Time with 1 to 9, Grades 6–12

Spend Unequal Fraction Time with 1 to 9

Note: In each problem the 1 digit must be used as the numerator for the fraction with a denominator of 2 in order to create only proper fractions.

a. $\dfrac{\boxed{1}}{2} < \dfrac{4}{\boxed{6}} < \dfrac{\boxed{7}}{8}$

Several other solutions are possible.

b. $\dfrac{\boxed{1}}{2} < \dfrac{\boxed{3}}{4} < \dfrac{8}{\boxed{9}}$

Only one possible solution.

c. $\dfrac{4}{\boxed{9}} < \dfrac{\boxed{1}}{2} < \dfrac{\boxed{5}}{8}$

The other possible solutions are the numerator for 8 can be 6 or 7.

d. $\dfrac{\boxed{}}{4} < \dfrac{\boxed{1}}{2} < \dfrac{8}{\boxed{}}$

Not possible to put a numerator over the 4 so that the value is less than 1/2.

e. $\dfrac{\boxed{3}}{8} < \dfrac{\boxed{1}}{2} < \dfrac{4}{\boxed{5}}$

The other possible solutions are the denominator for 4 can be 6 or 7.

f. $\dfrac{8}{\boxed{}} < \dfrac{\boxed{1}}{2} < \dfrac{\boxed{}}{4}$

Not possible to put a denominator below the 8 so that the value is less than 1/2.

© 2014 Consortium on Reaching Excellence in Education, Inc.

Spend Some Time with 1 to 9, Grades 6–12

Challenge 7 Solution

Multiply Your Fraction Time with 1 to 9

Some problems have more than one possible solution, although only one solution is shown below for each problem as an example.

1. $\dfrac{1}{2} \times \dfrac{3}{6} = \dfrac{\boxed{8} - \boxed{7}}{\boxed{4}}$

2. $\dfrac{1}{2} \times \dfrac{6}{3} = \dfrac{\boxed{4} + \boxed{8}}{\boxed{5} + \boxed{7}}$

3. $\dfrac{\boxed{2}}{\boxed{1}} \times \dfrac{\boxed{6}}{\boxed{3}} = \dfrac{4 + 7 + 9}{5}$

4. $\dfrac{\boxed{4}}{5} \times \dfrac{2}{\boxed{3}} = \dfrac{1 + 7}{\boxed{6} + \boxed{9}}$

5. $\dfrac{\boxed{3}}{\boxed{2}} \times \dfrac{1}{4} = \dfrac{\boxed{9} - \boxed{6}}{8}$

6. $\dfrac{\boxed{2} + \boxed{5}}{8} \times \dfrac{4}{6} = \dfrac{7}{\boxed{3} + \boxed{9}}$

7. $\dfrac{\boxed{4} + \boxed{5}}{6} \times \dfrac{1}{3} = \dfrac{\boxed{8}}{\boxed{7} + \boxed{9}}$

Challenge 8 Solution

Spend Some Time with 1 to 9, Grades 6–12

Divide a Fraction of Time with 1 to 9

1. $\dfrac{2}{3} \div \dfrac{1}{4} = \dfrac{\boxed{8}}{\boxed{9} - \boxed{6}}$

2. $\dfrac{2}{5} \div \dfrac{1}{3} = \dfrac{\boxed{6}}{\boxed{9} - \boxed{4}}$

3. $\dfrac{2}{3} \div \dfrac{6}{7} = \dfrac{\boxed{8} - \boxed{1}}{\boxed{4} + \boxed{5}}$

4. $\dfrac{1}{5} \div \dfrac{3}{4} = \dfrac{\boxed{6} - \boxed{2}}{\boxed{8} + \boxed{7}}$

5. $\dfrac{\boxed{9} - \boxed{5}}{6} \div \dfrac{1}{4} = \dfrac{8}{3}$

6. $\dfrac{\boxed{8} - \boxed{5}}{4} \div \dfrac{3}{\boxed{9} - \boxed{7}} = \dfrac{1}{2}$

7. $1\dfrac{2}{3} \div \dfrac{5}{9} = \boxed{7} - \boxed{4}$

© 2014 Consortium on Reaching Excellence in Education, Inc.

53

All Around and Solid with 1 to 9

Shaded surface areas (SA) identify correct answers.

Base	Height	Depth	SA
1	2	3	22
1	2	4	28
1	2	5	34
1	2	6	40
1	2	7	46
1	2	8	52
1	2	9	58
1	3	4	38
1	3	5	46
1	3	6	54
1	3	7	62
1	3	8	70
1	3	9	78
1	4	5	58
1	4	6	68
1	4	7	78
1	4	8	88
1	4	9	98
1	5	6	82
1	5	7	94
1	5	8	106
1	5	9	118
1	6	7	110
1	6	8	124
1	6	9	138
1	7	8	142
1	7	9	158
1	8	9	178

Base	Height	Depth	SA
2	3	4	52
2	3	5	62
2	3	6	72
2	3	7	82
2	3	8	92
2	3	9	102
2	4	5	76
2	4	6	88
2	4	7	100
2	4	8	112
2	4	9	124
2	5	6	104
2	5	7	118
2	5	8	132
2	5	9	146
2	6	7	136
2	6	8	152
2	6	9	168
2	7	8	172
2	7	9	190
2	8	9	212
3	4	5	94
3	4	6	108
3	4	7	122
3	4	8	136
3	4	9	150
3	5	6	126
3	5	7	142

Base	Height	Depth	SA
3	5	8	158
3	5	9	174
3	6	7	162
3	6	8	180
3	6	9	198
3	7	8	202
3	7	9	222
3	8	9	246
4	5	6	148
4	5	7	166
4	5	8	184
4	5	9	202
4	6	7	188
4	6	8	208
4	6	9	228
4	7	8	232
4	7	9	254
4	8	9	280
5	6	7	214
5	6	8	236
5	6	9	258
5	7	8	262
5	7	9	286
5	8	9	314
6	7	8	292
6	7	9	318
6	8	9	348
7	8	9	382

Challenge 10 Solution

Spend Some Time with 1 to 9, Grades 6–12

Create Equations with the Digits 1 to 9 and Their Opposites

There are many different possible equations. Here is an example of an equation that uses all 18 digits:

$$9 + -9 + 8 + -8 + 7 + -7 + 6 + -6 + 5 = 4 + -4 + 3 + -3 + 2 + -2 + 1 + -1 - -5$$

Spend Some Time with 1 to 9, Grades 6–12 **Challenge 11 Solution**

Spend a Proportional Amount of Time with 1 to 9

There are many possible answers, such as the following:

$$\frac{8}{6} = \frac{1}{3}\,(4) \qquad \frac{6}{3} = \frac{1}{4}\,(8) \qquad 2 = \frac{1}{4}\,(8) \qquad 8 = \frac{6}{3}\,(4)$$

Notice that for any proportional relationship, several more proportional relationships can be formed, for example:

$$8 = \frac{6}{3}\,(4) \quad \rightarrow \quad 4 = \frac{8}{6}\,(3) \quad \rightarrow \quad 3 = \frac{6}{8}\,(4) \quad \rightarrow \quad 6 = \frac{3}{4}\,(8) \quad \rightarrow \quad 6 = \frac{8}{4}\,(3) \text{ , etc.}$$

56   © 2014 Consortium on Reaching Excellence in Education, Inc.

Challenge 12 Solution

Spend Some Time with 1 to 9, Grades 6–12

Spend Proportional Time with Word Problems and 1 to 9

Answers will vary. Sample answers are shown below.

1. a. Jack runs 5 miles every 23 minutes. How long will it take him to run 1 mile?

    b. Virginia spends $16 every 4 days on transportation to work. How much does she spend in 8 days?

2. a. Jack runs 4 miles every 16 minutes. How long will it take him to run 8 miles?

    b. Virginia spends $18 every 3 days on transportation to work. How much does she spend in 7 days?

3. Many possible answers.

© 2014 Consortium on Reaching Excellence in Education, Inc.

Spend Some Time with 1 to 9, Grades 6–12

Challenge 13 Solution

Spend Exponential Time with 1 to 9

2. Problem d is not possible because for

$$4^{\boxed{}} < 8^{\boxed{}} < \boxed{}^2 \text{ to be true}, 8^{\boxed{}} < \boxed{}^2.$$

The exponent for 8 cannot be 3 or above, nor can it be 1.

- If 3 or more is the exponent for the 8, then there is no number that can be the base with exponent 2 that will work (For example, the greatest number that can be a base with exponent 2 is 9; however, $8^{\boxed{3}} < \boxed{9}^2$ is not true.)

- If 1 is the exponent with 8, then the smallest remaining possible value for the first term with the 4 is with an exponent of 3, but $4^{\boxed{3}} < 8^{\boxed{1}}$ is not true.

3. There are many possible answers for all but d and e. Two are shown with each statement to the right for a, b, c, and f.

4. Answers will vary.

a. $4^{\boxed{3}} < \boxed{9}^2 < 8^{\boxed{5}}$

$4^{\boxed{1}} < \boxed{5}^2 < 8^{\boxed{3}}$

b. $\boxed{3}^2 < 4^{\boxed{5}} < 8^{\boxed{7}}$

$\boxed{7}^2 < 4^{\boxed{3}} < 8^{\boxed{5}}$

c. $\boxed{5}^2 < 8^{\boxed{3}} < 4^{\boxed{7}}$

$\boxed{1}^2 < 8^{\boxed{3}} < 4^{\boxed{5}}$

d. $4^{\boxed{}} < 8^{\boxed{}} < \boxed{}^2$

Not possible.

e. $8^{\boxed{1}} < 4^{\boxed{3}} < \boxed{9}^2$

This is the only answer possible.

f. $8^{\boxed{1}} < \boxed{3}^2 < 4^{\boxed{5}}$

$8^{\boxed{1}} < \boxed{5}^2 < 4^{\boxed{3}}$

© 2014 Consortium on Reaching Excellence in Education, Inc.

Challenge 14 Solution

Spend Some Time with 1 to 9, Grades 6–12

Spend Some Radical Time with 1 to 9

2. Multiple answers are possible for each statement. Two are shown for each. (There are only two possible solutions for b and d. Both are shown.)

3. Problem d is not possible to solve.

    d. For $\boxed{}\sqrt{2} < \sqrt{4\boxed{}}$ to be true, only 1 and 3 can be used next to the radical 2. If 1 or 3 is used next to the radical 2, then there is no number that can be used in the first radical in the statement (next to the 8 under the radical sign) that will be less than the middle radical. $\sqrt{\boxed{}8} < \boxed{}\sqrt{2}$ will not be true if 1 or 3 is placed in the box next to the radical 2.

4. Answers will vary.

a. $\boxed{1}\sqrt{2} < \sqrt{4\boxed{3}} < \sqrt{5\boxed{8}}$

    $\boxed{3}\sqrt{2} < \sqrt{4\boxed{9}} < \sqrt{7\boxed{8}}$

b. $\sqrt{4\boxed{3}} < \boxed{5}\sqrt{2} < \sqrt{7\boxed{8}}$

    $\sqrt{4\boxed{1}} < \boxed{5}\sqrt{2} < \sqrt{9\boxed{8}}$

c. $\sqrt{\boxed{1}8} < \sqrt{4\boxed{9}} < \boxed{5}\sqrt{2}$

    $\sqrt{\boxed{3}8} < \sqrt{4\boxed{7}} < \boxed{5}\sqrt{2}$

d. *Not possible.*

e. $\sqrt{4\boxed{3}} < \sqrt{5\boxed{8}} < \boxed{7}\sqrt{2}$

    $\sqrt{4\boxed{5}} < \sqrt{7\boxed{8}} < \boxed{9}\sqrt{2}$

f. $\boxed{1}\sqrt{2} < \sqrt{3\boxed{8}} < \sqrt{4\boxed{5}}$

    $\boxed{1}\sqrt{2} < \sqrt{3\boxed{8}} < \sqrt{4\boxed{7}}$

© 2014 Consortium on Reaching Excellence in Education, Inc.

Spend More Radical Time with 1 to 9

2. Multiple answers are possible for each statement. Two are shown for each.

3. Only c is not possible.

   In order for $\sqrt{8\square} < 4\sqrt{\square}$ to be true, only the numbers 7 or 9 can be placed in the second radical (with the 4). However, if 7 or 9 is placed in this radical, then overall we have the radical values of

   $4\sqrt{\boxed{7}} = \sqrt{112}$ or $4\sqrt{\boxed{9}} = \sqrt{144}$.

   In either case, the final radical in the rightmost position, $\sqrt{\square}2$, must be more than the square root of 112, which is not possible since the greatest number we can have under this radical is 92.

4. Answers will vary.

a. $\sqrt{\boxed{1}2} < 4\sqrt{\boxed{3}} < \sqrt{8\boxed{5}}$

   $\sqrt{\boxed{3}2} < 4\sqrt{\boxed{5}} < \sqrt{8\boxed{7}}$

b. $4\sqrt{\boxed{1}} < \sqrt{\boxed{3}2} < \sqrt{8\boxed{5}}$

   $4\sqrt{\boxed{3}} < \sqrt{\boxed{5}2} < \sqrt{8\boxed{1}}$

c. $\sqrt{8\square} < 4\sqrt{\square} < \sqrt{\square 2}$

   Not possible.

d. $\sqrt{8\boxed{5}} < \sqrt{\boxed{9}2} < 4\sqrt{\boxed{7}}$

   $\sqrt{8\boxed{1}} < \sqrt{\boxed{9}2} < 4\sqrt{\boxed{7}}$

e. $4\sqrt{\boxed{3}} < \sqrt{8\boxed{1}} < \sqrt{\boxed{9}2}$

   $4\sqrt{\boxed{1}} < \sqrt{8\boxed{7}} < \sqrt{\boxed{9}2}$

f. $\sqrt{2\boxed{5}} < \sqrt{8\boxed{3}} < 4\sqrt{\boxed{7}}$

   $\sqrt{2\boxed{3}} < \sqrt{8\boxed{1}} < 4\sqrt{\boxed{9}}$

Challenge 16 Solution

Spend Some Time with 1 to 9, Grades 6–12

Make It Right with 1 to 9

Base	Alt.	Hyp.
1	2	$\sqrt{5}$
1	3	$\sqrt{10}$
1	4	$\sqrt{17}$
1	5	$\sqrt{26}$
1	6	$\sqrt{37}$
1	7	$\sqrt{50}$
1	7	$5\sqrt{2}$
1	8	$\sqrt{65}$
1	9	$\sqrt{82}$
2	3	$\sqrt{13}$
2	4	$\sqrt{20}$
2	4	$2\sqrt{5}$
2	5	$\sqrt{29}$
2	6	$\sqrt{40}$
2	6	$2\sqrt{10}$

Base	Alt.	Hyp.
2	7	$\sqrt{53}$
2	8	$\sqrt{68}$
2	8	$2\sqrt{17}$
2	9	$\sqrt{85}$
3	4	5
3	5	$\sqrt{34}$
3	6	$\sqrt{45}$
3	6	$3\sqrt{5}$
3	7	$\sqrt{58}$
3	8	$\sqrt{73}$
3	9	$\sqrt{90}$
3	9	$3\sqrt{10}$
4	5	$\sqrt{41}$
4	6	$\sqrt{52}$
4	6	$2\sqrt{13}$

Base	Alt.	Hyp.
4	7	$\sqrt{65}$
4	8	$\sqrt{80}$
4	8	$4\sqrt{5}$
4	9	$\sqrt{97}$
5	6	$\sqrt{61}$
5	7	$\sqrt{74}$
5	8	$\sqrt{89}$
5	9	$\sqrt{106}$
6	7	$\sqrt{85}$
6	8	10
6	9	$\sqrt{117}$
7	8	$\sqrt{113}$
7	9	$\sqrt{130}$
8	9	$\sqrt{145}$

© 2014 Consortium on Reaching Excellence in Education, Inc.

Spend Some Time with 1 to 9, Grades 6–12   **Challenge 17 Solution**

Fill In a Linear Pattern with 1 to 9

IN	OUT
2	3
4	7
8	15

Rule:
OUT = (IN × 2) − 1

IN	OUT
2	3
4	7
5	9

Rule:
OUT = (IN × 2) − 1

IN	OUT
2	1
5	7
6	9

Rule:
OUT = (IN × 2) − 3

IN	OUT
2	1
4	5
6	9

Rule:
OUT = (IN × 2) − 3

IN	OUT
3	6
4	9
5	12

Rule:
OUT = (IN × 3) − 3

IN	OUT
2	3
4	9
6	15

Rule:
OUT = (IN × 3) − 3

IN	OUT
1	5
2	7
3	9

Rule:
OUT = (IN × 2) + 3

IN	OUT
2	7
3	9
6	15

Rule:
OUT = (IN × 2) + 3

Challenge 18 Solution

Spend Some Time with 1 to 9, Grades 6–12

Make a Point of Slope with 1 to 9

1. There are many correct solutions, for example:

    $(7, 9)$ and $(6, 4)$ → $(9 − 4)/7 − 6) = 5/1$

    $(9, 6)$ and $(4, 7)$ → $(6 − 7)/9 − 4) = −1/5$

    $(9, 6)$ and $(2, 5)$ → $(6 − 5)/9 − 2) = 1/7$

    $(9, 5)$ and $(2, 1)$ → $(5 − 1)/(9 − 2) = 4/7$

2. There are many correct solutions, for example:

    $(19, 4)$ and $(25, 7)$ → $(7 − 4)/25 − 19) = 3/6$

    $(13, 5)$ and $(6, 9)$ → $(9 − 5)/(6 − 13) = 4/−7$

    $(1, 5)$ and $(8, 34)$ → $(34 − 5)/(8 − 1) = 29/7$

3. It is not possible to use only even or only odd digits because from 1 to 9 there are only five odd digits and only four even digits; however, to have two distinct points and a slope requires at least six digits.

© 2014 Consortium on Reaching Excellence in Education, Inc.

Spend Some Time with 1 to 9, Grades 6–12

Challenge 19 Solution

Make a Line with 1 to 9

1. Many solutions are possible, for example:

$$y = \frac{3}{4}x + 5 \quad \text{and} \quad y = \frac{6}{8}x + 9 \text{ ;} \qquad y = \frac{1}{3}x + 5 \quad \text{and} \quad y = \frac{2}{6}x + 7$$

2. Many solutions are possible, for example:

$$y = \frac{3}{4}x + 5 \quad \text{and} \quad y = -\frac{8}{6}x + 9 \text{ ;} \qquad y = -\frac{1}{3}x + 5 \quad \text{and} \quad y = \frac{6}{2}x + 7$$

64

© 2014 Consortium on Reaching Excellence in Education, Inc.

Challenge 20 Solution Spend Some Time with 1 to 9, Grades 6–12

Attack the Quadratic with 1 to 9

The tables below show the values used for *m* and *n*, and resulting values for *b* and *c* when the equation is rewritten in standard form. Shading shows the solutions.

m	n	b	c
1	2	3	2
1	3	4	3
1	4	5	4
1	5	6	5
1	6	7	6
1	7	8	7
1	8	9	8
1	9	10	9
2	3	5	6
2	4	6	8
2	5	7	10
2	6	8	12
2	7	9	14
2	8	10	16
2	9	11	18

m	n	b	c
3	4	7	12
3	5	8	15
3	6	9	18
3	7	10	21
3	8	11	24
3	9	12	27
4	5	9	20
4	6	10	24
4	7	11	28
4	8	12	32
4	9	13	36

m	n	b	c
5	6	11	30
5	7	12	35
5	8	13	40
5	9	14	45
6	7	13	42
6	8	14	48
6	9	15	54
7	8	15	56
7	9	16	63
8	9	17	72

© 2014 Consortium on Reaching Excellence in Education, Inc.